# Crewed Space Stations

by Patrick H. Stakem
(c) 2017, 2022

Number 9 in the Space Series

# Table of Contents

Introduction..................................................................................4
Author..........................................................................................5
A note on Units...........................................................................7
The early dreams, von Braun, Willy Ley, Tsiolkovsky...............8
    Manned Orbiting Lab (MOL)................................................10
    Skylab.....................................................................................11
    Skylab-II ................................................................................15
    Space Station Freedom...........................................................15
    International Space Station – ISS..........................................16
    Modules..................................................................................17
        ESA....................................................................................19
        Japan..................................................................................20
    Electrical Power.....................................................................21
    Static and "grounding"..........................................................22
    Environmental ......................................................................22
    Computers and communications............................................23
    Construction..........................................................................27
    Repairs...................................................................................27
    Operations.............................................................................28
    Logistics................................................................................30
    Decommissioning..................................................................31
    X-38 Crew Return Vehicle...................................................31
Planned Follow-ons...................................................................33
    Deep Space Gateway ............................................................33
    Deep Space Habitat...............................................................35
    Russian Lunar Orbital Station ..............................................37
    Chinese Lunar Exploration Program.....................................37
    LOP-G....................................................................................37
    NASA Lunar Outpost............................................................38
Commercial Efforts...................................................................39
    Blue Origin............................................................................39
    Bigelow Aerospace Commercial Space Station ..................41
    Orion Span............................................................................42
    Orbital Technologies Commercial Space Station.................43
    Axion Space..........................................................................43

- SpaceX......43
- Soviet Union/Russia......44
  - Salyut......45
  - Almaz......46
  - MIR......46
  - Shuttle-Mir Program......50
  - Opsek......51
- Chinese Tiangong......52
- India......53
- Afterthought......54
- Bibliography......55
- Resources......62
- Glossary of Terms ......64
- If you enjoyed this book, you might also be interested in some of these......72

"We choose to go to the moon in this decade and do the other things, not because they are easy, but because they are hard..." John F. Kennedy

"Mankind will not remain on the earth forever, but, in search of light and space, will at first timidly penetrate beyond the limits of the atmosphere and then finally conquer the spaces of the solar system.," Konstantin Tsiolkovsky's tombstone.

## Introduction

This book covers the topic of Crewed Space Stations, from the earliest dreams to the current International Space Station, with some information on the planning of its replacement. Generally, we distinguish between a space capsule and a space station by the fact that the space station is permanently in orbit, and can be resupplied and new crews delivered, as veterans are returned to the ground. The orbital space station provides living accommodations for the crew, as well as experiment space.

The first mention of a crewed space station may be Edward Everett Hale's *The Brick Moon*, in 1899. The first instance of the torus or wheel shape is usually attributed to Potocnic, an Austrian, in *The Problem of Space Travel,* published around 1928. Pirquet picked up and expanded on this concept in his book, *Die Rakette*, 1928. Hermann Noordung discussed space stations in his 1929 book, *The Problem of Space Travel: The Rocket Motor.* An orbital weapons station was researched in Germany during World War-II. It would have used directed solar energy.

Von Braun, by now in the U. S. popularized the Space Station idea with an article in *Collier's Weekly* in 1951. In the Soviet Union at the same time, orbital stations concepts from Tsiolkovsky led to the Almaz, Salyut, and Mir. Usually Space Stations are built in orbit, and manned by a crew when ready. They are usually modular, and are supplied by logistics flights. The current thinking is that crew duration in Space should be limited to about a year, so crew exchange is a regular feature. The Station hardware is designed to stay in Space, but has a definite operational lifetime.

The topic of Spacecraft to get humans to space (and back) is the subject of a companion volume. This book will discuss the options to deliver new crew and logistics supplies to existing and planned stations.

## Author

Mr. Patrick H. Stakem has been fascinated by the space program since the Vanguard launches in 1957. He received a Bachelors degree in Electrical Engineering from Carnegie-Mellon University, and Masters Degrees in Physics and Computer Science from the Johns Hopkins University. At Carnegie, he worked with a group of undergraduate students to re-assemble, modify, and operate a surplus missile guidance computer, which was later donated to the Smithsonian. He was brought up in the mainframe era, and was taught to never trust a computer you could lift.

He began his career in Aerospace with Fairchild Industries on the ATS-6 (Applications Technology

Satellite-6) program, a communication satellite that developed much of the technology for the TDRSS (Tracking and Data Relay Satellite System). He followed the ATS-6 Program through its operational phase, and worked on other projects at NASA's Goddard Space Flight Center including the Hubble Space Telescope, the International Ultraviolet Explorer (IUE), the Solar Maximum Mission (SMM), some of the Landsat missions, and Shuttle. He was posted to NASA's Jet Propulsion Laboratory for Mars-Jupiter-Saturn (MJS-77), which later became the *Voyager* mission, and which is still operating and returning data from outside the solar system at this writing. He initiated and lead the international Flight Linux Project for NASA's Earth Sciences Technology Office. He is the recipient of the Shuttle Program Manager's Commendation Award, and has completed 42 NASA Certification courses. He has two NASA Group Achievement Awards, and the Apollo-Soyuz Test Program Award.

Mr. Stakem was affiliated with the Whiting School of Engineering of the Johns Hopkins University since 2007. He supported the Summer Engineering Bootcamp Projects at Goddard Space Flight Center for 2 years. These resulted in the Greenland Rover, a tracked robot measuring the depth of the ice sheet.

Mr. Patrick H. Stakem has been fascinated by the space program since the Vanguard launches in 1957. He received a Bachelors degree in Electrical Engineering from Carnegie-Mellon University, and Masters Degrees in Physics and Computer Science from the Johns Hopkins University. At Carnegie, he worked with a group of undergraduate students to re-assemble, modify, and operate a surplus missile guidance computer, which was later donated to the Smithsonian. He was brought up in the mainframe era, and was taught to never trust a computer you could lift.

Now he supports International collaborative open-source projects, and STEM Programs.

Mr. Stakem can be found on facebook and LinkedIn. Comments, corrections, suggestions are appreciated.

All material is based on ITAR compliant sources.

## A note on Units

I am fairly conversant in both English and Metric units (what is the metric equivalent of furlongs per fortnight?). Metric (SI) is mandated for NASA usage now, for interchangeability with our partner space faring nations. When a lot of the legacy flights discussed here were flown, English units were the norm. I have tried to keep the units comparable to the mission at the time. Conversions are easy enough, but units conversion is a source of error. It's not what you know about units and measurement, its how you think. And, I still think English units (even the English use Metric now), and convert in my head or on my phone.

For scientific/engineering work, the Metric system is well thought out. For artisans, the English system served well, as most units were divided by 2. Which is easy. Fold the cloth. Hopefully, when we are all taught Metric first, some one will still remember the conversions. You just need a slide rule....

## The early dreams, von Braun, Willy Ley, Tsiolkovsky

Willy Ley and Werner von Braun in Germany were discussing space stations in the 1930's in Germany, as members of the German Rocket Society (VfR). In 1903, Konstantin Tsiolkovsky in Russia showed how to generate simulated gravity with the wheel or torus shape.

I met Willy Ley in the '60's at a Model Rocket meet. I had read his books when I was in grade school, and he launched one of my rockets. He is one of my true Hero's and a major driving force that got us into space. Unfortunately, he did not live to see the first lunar landing.

In the Introduction to Willy Ley's book, *For your Information*, Robert M. Guinn of Galaxy Publishing Corporation says of Dr. Ley that his publishing contract with Galaxy contained a termination clause, "if and when Willy Ley is assigned the job of placing a space station in orbit by the United States Government."

von Braun and Ley advocated the rotating torus for a Mars mission. With a diameter of 76 meters, if the wheel rotated at 3 rpm, the crew would experience 1/3 g. Author Arthur C. Clarke used a rotating space station in his novel, 2001: A Space Odyssey. In 1975, a NASA study at Sanford University came up with the Stanford Torus, a 1.8 km diameter habitat for 10,000 persons in orbit, rotating at 1 rpm. It was to be located at the Earth-Moon L1 Lagrangian point, where the gravity from the two bodies cancel each other.

I was working for Fairchild Aerospace in the 1970's when von Braun retired from NASA, and went to Fairchild as Vice President of Engineering. So, technically, I was briefly a member of the von Braun team.

The early concepts of a space station were torus-shaped. This facility would revolve around a central axis, and provide a level of gravity to the inhabitants based on the centripetal force of rotation. You could simulate Earth surface weight, but would probably just go with enough simulated gravity to feel comfortable and keep things in place. At the time, there was no data on whether humans could function in a zero-gravity environment. Now that we know they can, the space station designs can be more functional. Von Braun had envisioned a 250 foot diameter torus that rotated, providing near Earth gravity. He also anticipated a fleet of Shuttles for construction in orbit. He saw the facility as a jump-off point for lunar and Mars missions. We now know from experience that astronauts operate perfectly well in zero-G.

Konstantin Eduardovich Tsiolkovsky, of the Soviet Union, is considered to be another of the founding fathers of modern space hardware. He is still influencing the Russian space program to this day.

This section discusses the Soviet/Russian, American, Chinese, and other nations' efforts, including the current International Space Station (ISS) in which 15 nations participate. None of these follow the torus design.

## Manned Orbiting Lab (MOL)

The Manned Orbiting Lab was a USAF space program circa 1963. It was to be launched into orbit, and then two astronauts could visit it via a Gemini Capsule. Mission duration of up to 40 days were envisioned. The project was a follow on to Dynasoar, a winged space plane.. Hardware was built, but the mission was never flown. There were to be seven flights, 5 of them crewed. MOL astronauts in three groups were chosen from Air Force, Navy, and Marine Corps personnel. The project was canceled in 1969. NASA's Skylab mission was the follow-on to a lab in orbit. A Gemini-B Capsule for MOL is at the U.S. Air Force Museum, Dayton, Ohio It has a circular hatch in the heat shield for entering the lab.

MOL would have been launched on an upgraded Titan vehicle. The capsule would dock with the MOL unit in-orbit, and at the end of the mission, would undock, and return to Earth. It was designed to operated in Polar orbit, which would involve a launch from Vandenberg Air Force base in California. The reaction control system for attitude control in orbit was different from the Gemini's OAMS.

A MOL mock-up and a refurbished Gemini capsule (not crewed) were launched from Cape Canaveral on a Titan-IIIc in 1966. The Gemini capsule reentered and was recovered. The MOL released three satellites, once it achieved orbit. Proposed first crewed flight was to be in 1970, but the program was canceled. NASA offered the military crews the opportunity to transition to the civilian space corps, and the 7 that were eligible did so. All went on to fly on shuttle missions.

The MOL was to be built by Douglas Aircraft. The project was canceled, since it had been demonstrated that un-crewed reconnaissance satellites could spy on our enemies much more cost-effectively.

## Skylab

After the Apollo program, with some spare Saturn's sitting around, the next project was the Skylab space station. This used a Saturn S-IVB upper stage as the structure for the station, launched by a Saturn-V with live first and second stages. The hydrogen fuel tank was repurposed into the crewed facility. The payload to orbit was 170,000 pounds. The station was 82 feet long, 56 feet wide, and 36 feet in the other direction. It was quite visible from Earth, when the solar arrays caught the glint of the Sun. Astronauts were carried to the facility in-orbit on three missions in 1973-1974 by Apollo command and service modules launched on Saturn-Ib vehicles. A second Saturn-1b and Apollo stack was kept in standby in case a rescue mission was needed. The lab's orbit was a near-circular 270 miles, with a 50 degree inclination. It was competing for funds and resources with the USAF's MOL project at the time.

The facility had been damaged during launch when the integral micrometeriod shield torn away, taking one of the solar array panels with it. This caused thermal and power problems. During an un-anticipated EVA, the crew rigged a replacement heat shield, and freed the solar panel. The third expedition set a record for days spent in space, at that time, 84 days.

The Industrial design firm for Skylab was headed by famed architect Raymond Lowey. He emphasized habitability and comfort for the astronaut, He included a wardroom space for meals and relaxation. He also wanted a window to view Earth and space. This has proved to be a great feature on the current ISS. Astronauts who participated in Skylab planning were dubious about the designers' focus on areas such as color schemes and decoration. They vetoed an entertainment center. Skylab food was improved over the earlier lunar mission food.

With a large volume of science data to be managed, the Apollo Guidance Computer (AGC) was not up to the job. The IBM System/4Pi TC-1, a derivative of the IBM S/360 mainframe, and a relative of the subsequent Shuttle's AP-101, was used. This was a radiation-hardened unit, with a 16-bit word, and 16k of memory. It had a custom-designed input-output unit for the lab. It drew 56 watts of electrical power, and weighed 18 pounds. It was built with ttl-technology integrated circuits, and core memory. Ten were built, and two were flown. A standard S/360 mainframe was used to produce code and a simulator was used for verification. An IBM System/360-75 mainframe was used to evaluate the onboard computer's performance in orbit. It could run a simulation at 3.5 times less than real-time.

The 4Pi had 54 different instruction, and supported 32-bit double precision data. Cycle time was 3 microseconds. An add took 3 cycles, and a multiply or divide took 16. It had a 24-bit real-time clock. A simple numeric keyboard was used to communicate with the computer in the lab in orbit, and it could also be loaded via uplink, and by an onboard tape drive.

Liquid waste was not recycled in Skylab as it is on the current ISS. Liquid and solid waste went into the large Oxygen tank in the facility, and was stored. It burned when Skylab re-entered.

The facility provided living and working space for the astronauts, a true shower and toilet, a solar observatory, and fixtures and services for science experiments. Some experimental data including film was brought back with the astronauts. The facility had two docking ports, and an airlock. When an Apollo capsule was docked, electrical power from the fuel cells in the service module could be used to augment power from the solar arrays.

The first Skylab mission lasted 272 days, not all occupied, followed by an idle period of 394 days, when the computer kept things going. The computer was turned off for 4 years while NASA discussed reboosting Skylab to a higher orbit, or letting it reenter. There was a need to put some mods in the software, but the tools and card decks containing the code had been discarded. This resulted in some 2500 cards being re-punched from code listings. At the end of 4 years, the onboard computer was booted up by ground command, and the updates worked fine.

There were plans to use the Shuttle to repair and reboost Skylab, but the timing did not work out. Skylab was in orbit until 1979, when it reentered the atmosphere, splashing into the Pacific ocean near Perth, Australia.

There was also an entire second Skylab spare. The Skylab 2 / Crew 1 Command Module is at the Naval Aviation Museum, Pensacola, Florida. Skylab 3 / Crew 2 Command Module is at the NASA Visitor Center at Great Lakes Science Center, Cleveland, Ohio. Skylab 4 / Crew 3 Command Module is at the National Air and Space Museum, Washington, D.C. The service modules were jettisoned before the capsules reentered the atmosphere.

The Skylab at the Smithsonian is the functional flight spare unit. The active unit reentered the atmosphere in 1979. Two pieces were recovered and are on display. The museum in Esperance, Western Australia, has some pieces, as does the Alabama Space & Rocket Center Museum.

As follow-ons to Skylab, a number of proposals were presented. Von Braun envisioned a larger station, built from the second stage of the Saturn-V. The second stage would be used for fuel and liquid oxygen on the way to orbit, then vented to space, and an equipment module would be slide into the vented hydrogen tank. Questions about how many Saturn-V's would be available led to the choice of the S-IVB option.

## Skylab-II

Skylab-II was a circa 2013 concept from the Marshall Space Flight Center's Advanced Concepts Office. It would be the same concept as the original Skylab, but used the upper stage hydrogen fuel tank from the Space Launch System then under development. It is to be located at the Earth-Moon L2 point (Lagrange point, a null in the gravity field). Here, it would need minimal orbital adjust to remain at that spot. That particular point is on the other side of the moon, from the Earth. That puts it some 430,000 km from Earth, and 62,800 km from the lunar surface. With the moon between the station and the Earth, a relatively quiet radio environment is achieved. The goal is to support a 4 person crew for 60 days, without a resupply flight. The re-purposed tank would have a diameter of 8.5 meters, larger than the ISS's 4.5 meters. The provided volume would be about 500 cubic meters. Lessons learned from the ongoing ISS mission would be applied to the Skylab-II project.

## Space Station Freedom

Space Station Freedom was to the the U.S. follow-on to Skylab in the 1980's. At that time, Russia was working on their third version of the Saylut program, named Kosmos 557. This suffered a failure in orbit and was allowed to reenter. Cost issues for both parties brought the Americans and Russians together on a joint venture, to be called the International Space Station. Many of the concepts and lessons-learned on-orbit were incorporated into the new design.

## International Space Station – ISS

In 1993, United State's Space Station Freedom Project, to create the International Space Station kicked off. On-orbit construction began in 1998, and was completed with a last Shuttle mission in 2011. It is the largest artificial satellite in Earth orbit, and can be seen from the ground with the naked eye. The ISS is a synthesis of several space station modules from the U. S., the Soviets/Russians, the Europeans, and the Japanese. It serves as a laboratory, observatory, and factory in Earth orbit, and is continuously crewed. Part of its mission is to collect information on items in orbit for long duration. It is currently funded through 2024. The assembly began in 1998, with the first module being the Zarya, It now has 15 pressurized modules. Five more are planned,. This is the advantage of a modular architecture. It is the 11$^{th}$ station sent to orbit. Early stations such as Skylab were not intended for resupply.

The International Space Station is continuously crewed, and orbits the Earth at an altitude of some 250 miles. It is quick, traveling at 17,300 miles per hour. It is also expensive, representing an investment of some $100+ billion dollars by the world community, mostly by the United States and Russia. It is thus the most expensive object ever constructed by mankind. It has been visited by astronauts and cosmonauts from some 15 nations, and by paying tourists. It can generally be found at an altitude between 300 and 435 km, and can be seen by the naked eye in the daytime, if you know here to look (there's a NASA ap for that).It has be continuously occupied for 16 years, as of this writing. It has been visited by travelers from 17 nations, some for work, some for tourism. It normally has a crew of 6, and masses 419,500 kg, the largest item in orbit.

It is a joint project with Russia, but they indicated leaving the program in 2024, to build their next-gen station, The Russian Orbital Service Station.

# Modules

The modules include Zarya (Russian, "dawn") which is a functional cargo block, and the first to be launched. It was built in Russia, but belongs to the U. S. The Unity module is a passive connecting module, and was brought up on the Shuttle Endeavor. The Quest module houses Russian and US spacesuits for EVA's. It is also used as a sleeping compartment. Zveda (Russian, "star") is a service module. It provides life support for 6 crew, and sleeping quarters for two. The Destiny modules is the primary research lab facility for the U. S. PIRS and Poisk are Russian docking modules. Due to a fault encountered on MIR, all ISS hatches were designed to open inward. Harmony is a node module, providing power and data to the modules connected to it, the European Columbus, and the Japanese Kibo labs. It is also a docking port, supporting the Japanese HTV, Space-X Dragon, and the Orbital-ATK Cygnus. Tranquility is a US module, providing life support functions. All US modules have 6 berthing adapters, to connect to other modules. It is currently connected to the Station core, and hosts the Cupola, a docking port, the Bigelow commercial "Expandable Activity Module," and Leonardo, a European storage module built in Italy.

The Cupola is an observatory with 7 windows. It was a NASA-ESA Project, built in Turin, Italy. The Rassvet ("dawn") module is the Russian mini-research module. It was originally a docking module.

Columbus is the European research facility, with provisions for external experiments. Kibo is the Japanese laboratory, the largest module on the ISS. It includes an airlock with a drawer, to deploy payloads in the space environment.

The current PIRS module is scheduled to be de-orbited. PiRS is a docking compartment; the word means "pier" in Russian. Two other modules, Zarya, currently owned by NASA, and Rassvet, docked at Zarya, will be replaced.

The Bigelow Expandable Activity Module (BEAM) is a privately owned, inflatable habitat module, developed from NASA technology. It came to the station in 2016, and, so far, has remained inflated. Bigelow Aerospace hopes to use such modules in its own space habitat project. In more than a year in orbit, it has worked well. It is currently used as a storage area.

The Nauka (Russian: "science") module is known as the Multipurpose Laboratory Module The Uzlovoy (Russian "masculine") module is a ball-shaped docking module that has not yet been launched. It is supposed to be placed on the Earth-facing side of Nauka, and has 6 docking ports.

An ISS model is at the National Air and Space Museum in Washington, D. C. The station is huge, and designed to operate in zero gravity, so a complete full scale model is not feasible. There is a small model at the Goddard Space flight Center Visitor's Center in Greenbelt, Maryland. The best exhibit is at the Marshall Space flight Center, in Huntsville, AL.

ESA

The European Space Agency contributed a laboratory module to the ISS, called *Columbus*. It was built by Thales Alenia Space, in Turin, Italy, and was delivered by the Shuttle Atlantis in 2008. It is controlled from the German Space Operations Center in Munich, Germany. It is a cylindrical module, 23 feet long, with two end cones. There is a berthing mechanism at one end. The module contains ten standard payload racks (ISPR), of which 5 are used by ESA, and 5 by NASA. Unpressurized payload platforms are available at the Columbus external payload facility. The current Columbus project was a downgrading of an earlier ESA project for a crewed space station of their own.

Japan

The Japanese Experiment Module (JEM, named Kibo) is the largest on the station. It came up on three Shuttle missions. There are six major elements, including the pressurized lab,and the exposed facility, and there is a robot arm. The pressurized module is used for press conferences.

Besides the various pressurized modules of the station, there is considerable space outside (no pun intended). The station's Integrated Truss Structure is the backbone. It is nearly 110 meters in length, The main solar arrays and thermal radiators are attached to this truss. The solar cells are of a special design, bifacial. They have two generating surfaces. The solar arrays can be aligned to the Sun, and the other side will receive enough reflected sunlight from Earth to contribute. The station currently has 6 robot arms as well as a mobile crane. The Mobile Servicing System consists of Candarm2, the Mobile Remote Servicer Base, and the Special Purpose Dexterous Manipulator. With the Shuttle-derived latching end effector, the system can latch onto and move large assemblies out the station. In addition, the entire assembly is mounted to the Mobile transporter cart, which rides on rails on the Station's Integrated truss structure.

The Russian NEM-1 and NEM-2 modules are for power for science instruments. They have not been launched yet, as of this writing.

The station also includes a NanoRacks airlock module, which is used to deploy Cubesats from the station. A Cubesat is a small, affordable satellite that can be developed and launched by college, high schools, and even individuals. The specifications were developed by Academia in 1999. The basic structure is a 10 centimeter cube, (volume of 1 liter) weighing less than 1.33 kilograms. This allows multiples of these standardized packages to be launched as secondary payloads on other missions. A Cubesat dispenser has been developed, the Poly-PicoSat Orbital Deployer, P-POD, that holds multiple Cubesats and dispenses them on orbit. Cubesats sent to the ISS can be returned to the ground on subsequent return of logistics carriers.

# Electrical Power

Power on the station comes from the driven solar arrays, which track the Sun. Most of the equipment uses the 28 volt dc standard. The US module stabilizes the array power to 160 volts-dc for distribution, and to 124 vdc for use. There is a trade-off here. The higher the voltage, the smaller the conducting wire can be. If you grab a hot 28 volt wire, you won't notice anything. But more than 100 volts can ruin your day.

The arrays can produce power for about 60 minutes of every 90 minutes, the orbital period. Think of a 1 hour "day", followed by a 35 minute "night." Rechargeable nickel-hydrogen battery's are used for the dark period. Charge-discharge cycles are monitored, and their life is around 6.5 years, or 36,000 charge-discharge cycles. These are being replaced by lithium-ion units, with longer lifetimes, and better energy density.

## Static and "grounding"

One problem the station has is static charge. The Spacecraft charging phenomenon has been known for decades. The issue is that every thing has to be "grounded" to everything else (conductive paths), to avoid a spark, which will damage the electronics. Static potentials of thousands of volts can accumulate. On Earth, we would make sure everything was grounded. In space, there is no "ground." But the Station is surrounded by a thin plasma field at that altitude, and special plasma contactor units "ground" the ISS to that potential. It's the differences in voltage that are the problem.

## Environmental

The station interior environment is regulated for crew health and comfort. There is an excess of heat generated by equipment and crew, and this is collected by a system of cooled ammonia circulated through pipes in the modules. Radiators on the main truss dump this heat to space. There is also a passive thermal control system, as the modules are wrapped in multilayer insulation, to prevent being heated by direct radiation from the Sun. The active heat removal system has a capacity of 70 kilowatts.

The Station atmosphere is kept at a pressure and composition similar to sea-level. Oxygen is generated onboard to replenish that used by the astronauts, and carbon dioxide is removed. The air is filtered to remove unwanted particles and chemicals. There are three oxygen supplies on the station. First, there is a supply of bottled, pressurized gas. Then, there are the solid fuel generators, that generate excess oxygen in a combustion process. Thirdly, a unit can be used to electrolyze water, and vent the hydrogen. This unit requires about a liter of water a day, for each crew member. Recycled water on the station is used.

The station is kept at zero gravity, a condition that has been proven to be feasible for crew staying up to a year in space. That's about the limit for radiation exposure. Before any person was sent to orbit, it was though that some form of artificial gravity would need to be provided for the crew. Turns out, they exercise daily, and are fine.

# Computers and communications

The station has a complex data and computing infrastructure. The onboard computing was based on rad-hard Intel 80386's for housekeeping., environmental monitoring and control, and station-keeping (orbital maneuvers).

There is an Ethernet network backbone in the Station, as

well as WiFi. Applications include IP phone with webcam for crew conversations with familys.

The laptops replaced the original design of the Multi-Purpose Application consoles, which were MIL-Spec equipment, purpose-built and programmed from scratch using the ADA programming language and X-windows. The laptops (PGSC – Portable General Support Computers) were more flexible. The general concept is that COTS machines can display any data, but must follow a arm/check/fire protocol to send commands. Originally, the support computers were Grid laptops. These were 8086-based machines, running GridOS.

The Station uses some 68 COTS IBM/Lenovo ThinkPad A31 laptops, 32 ThinkPad T61 laptops, and a T61p as a server, with routers. The laptops handle non-critical and experimental support as well as music and television for crew entertainment. The computers ran Windows-XP, with some 3 million lines of flight code. The architecture was designed as a station-wide distributed system.

The laptops were modified to handle the environment of the station. With no convention cooling, due to lack of gravity, fans had to be added to keep components on the motherboard cool. In addition, the circuit boards and connectors were conformally coated to contain any debris. The power supplies were modified to accept Station 28 vdc power. In addition, the laptop had to operate in the 10 psia atmosphere of the Space Shuttle, even though the Station maintains a sea-level pressure of

14.7 psia. The laptops had Velcro for attaching to convenient surfaces. The Laptops use the station wifi, which is linked to the ground via Ku band transceivers. It's 10 megabits up, 3 megabits down, sort of a space-equivalent DSL.

The latest addition to the ISS computer network is a pair supercomputer, It went up in a Dragon capsule, launched by a Falcon-9. It is a COTS Hewlett Packard Enterprise Apollo, with high speed interconnect. It runs the Linux operating system, an achieves a performance of 1 teraflop ($10^{12}$ floating point operations per second). A twin is kept on the running on the ground.

It is the pizza box server form factor. There are two racks of servers, and include X86 architecture cpu's with solid state storage drives. It is water-cooled, and the power supply was modified to use station power. The running software will monitor the computer's operation to check for the symptoms of radiation damage, which normally manifests itself as an increase in current draw. This can sometimes be overcome with a reboot.

The machine has several different purposes. First, it will show if a computer in space, normally subject to radiation induced upsets, can survive within the crewed space of the ISS and operate for a year. That is about the limit for crews. The second reason for the installation is to "take the computer to the data." It will crunch experiment data and allow downlinking of processed information instead of the raw data, vastly increasing the required bandwidth.

In addition, HP announced that NASA is going to Phase-in second generations laptops on the station, the Zbook 15 mobile workstation. These will replace the HP Z-series 8570's now in use. There are currently 100 laptops and workstations onboard.

The Russian orbital segment communicates with the ground directly, and can also use the Russian relay satellites. They have an internal telecommunications system, that can include a VHF link to the ground.

The Russian Orbital segment houses the Data Management System, which oversees Guidance, Navigation, and Control for the station. This is implemented in the ESA DMS-R computer. This consists of dual fault tolerant computers, each with triple modular redundancy. It interfaces with Earth horizon sensors, solar horizon sensors, and star trackers. For attitude control, it has gyroscopes and thrusters. A docked Soyuz can also maintain the station attitude for a while.

The US segments use an S-band and Ku band link to the ground via the TDRSS satellite relay. The Canadarm, the European Module, and the Japanese module have their own communications capability with their respective control centers throughout the station, there is an internal wireless network. For crew doing EVA operations, a UHF radio link is used. Spacecraft docking to the station also use a UHF link.

There is an amateur radio transceiver on the ISS, using 145.8 MHz for transmission, and 144.49 or 145.20 Mhz for receiving. Talks to school kids are commonly set up.

The ISS Android ap from NASA is great. It shows the point on the Earth the Station is above at the time, and has a high resolution camera that returns images (during Sun-lit portions of the orbit). Download it to your phone.

# Construction

The ISS was constructed in orbit, using large modular sections delivered to orbit by the Russian Proton heavy lift vehicle, and the Shuttle. Both had a capacity of around 50,000 pounds to Low Earth Orbit. The first section in orbit was the Zarya module.

Thirty-five Shuttle flights were used during the construction phase, to deliver components and construction crews, and serve as a temporary habitat for the crew to use.

# Repairs

Astronauts and Cosmonauts can go out through airlocks to effect repairs to the station. This is usually hazardous. In one case, a solar array had torn. The EVA crew had to work on it when it was in sunlight, which introduced an electrocution hazard (which the suit could not have handled). The repairs were carried out carefully and successfully. In another incident, the ammonia cooling system was damaged, and one of the Astronauts sent out to evaluate and fix it got covered in ammonia, which promptly froze to his suit. In this case, he had to wait until it sublimated away (and hope his oxygen held out), because he would have contaminated the airlock with the ammonia.

## Operations

The crews are rotated in and out of the station for varying periods, not to exceed a year. This service uses a Russian Soyuz capsule. There is always one capsule at the station for emergency evacuation. Each crew is referred to as an "Expedition." Current maximum is a crew of six. When a crew exchange does not involve all 3, (the capacity of the Soyuz) the spare seat may be sold to a "spaceflight participant" or space tourist. The current seat price is $40 million, includes room and board, return trip, and magnificent views. Seven have taken advantage of this opportunity to date. Cubesats coming up as cargo are accommodated by the NanoRacks system in the station, and can be attached to the outside, later to be retrieved and returned to the ground.

Logistics flights (food, oxygen science projects, clean underware) up, and trash down use a variety of options, the Soyuz cargo version, and commercial un-crewed capsules under contract to NASA. Some capsules burn in the atmosphere upon reentry, and some can be reused. Resupply is by Russia's *Progress* (5,200 lbs), Orbital-ATK's *Cygnus (*4400-7700 lbs, depending on launch vehicle*)*, and SpaceX's *Dragon* (6400 lbs), *t*he European Automated Transfer Vehicle (19,500 lbs) and the Japanese HTV Transfer vehicle (9900 lbs).

Due to orbital drag, the station needs occasional reboosting. This can be done with any capsule docked to the Russian service module Zvezda's aft port.

For sleeping arrangements, there are no dedicated sleeping modules, and crew just typically Velcro themselves to a wall in a module of their choice. There are two Waste and Hygiene Facilities of Russian design. The liquid is piped to the Water Recovery System, and solid waste is bagged and put in a returning Progress logistics carrier.

In normal operations, the Earth's magnetic field deflects charged particles from the station. Energetic space particles may pass through the station with negligible effect. Space debris is a problem, from discarded bolts to Zombie-Sats, dead satellites in orbit. These are all tracked, and on rare occasion, the station needs to do a damage avoidance maneuver to avoid a collision. Solar flares, Coronal Mass Ejections can also endanger the crew. They then have to gather in their equivalent of a tornado shelter. A series of sentinel satellites track these events originating in the Sun, and provide ample warning time. To date, no evacuation of the Station has been necessary. The Crew normally is exposed to the same radiation in 1 day, which a person one Earth would get in a year.

And, the answer as to whether there is life in space is

resoundingly "yes." By 2012, 76 types of microorganisms were detected on the station.

# Logistics

There are two main items that go up to and come down from the on-orbit ISS. Fresh crews replace space veterans, and supplies of food, clothing, science equipment, and toilet paper go up. This is done with an automated, non-crewed logistics vehicle. Trash comes down, with the simple yet costly expedient of having the carrier burn up in the atmosphere on reentry. Now that the Shuttle fleet has been decommissioned after two disasters, the U.S. currently has no human-rated logistics vehicle, and relies on the Russian Soyuz craft for crew exchange. Crew exchange is done currently only by the Soyuz TM, but several options are in work for a new series of U. S. crewed vehicles. Several commercial companies provide non-crewed logistics services, including Space-x (Dragon),

When the Shuttle program was still active, The Shuttle could stay docked to the iSS for a maximum of 12 days. The Soyuz crewed capsules can stay 180 days.

Access to the facilities of the ISS are defined in the Space Station Intergovernmental Agreement (IGA). This international treaty was signed on 28 January 1998 by the United States of America, Russia, Japan, Canada and eleven member states of the European Space Agency (Belgium, Denmark, France, Germany, Italy, The Netherlands, Norway, Spain, Sweden, Switzerland, and the United Kingdom).

# Decommissioning

The current ISS will reach end-of-life in the 2020's, and is too big to be allowed to re-enter in one piece. One or more follow-on stations will be built in orbit, re-using some modules from the ISS, and new modules launched from the ground. This is possible due to the modular nature of the ISS, and the lessons-learned during its construction and use.

Russia currently has plans to remove some of its modules, and re-purpose them into a new facility, the Orbital Piloted Assembly and Experiment Complex (OPSEK). This is based on an estimated life on-orbit of 30 years, from the MIR experience.

At the same time, the various nations that own parts of the station are responsible for their disposal. It the parts can not be re-purposed, they will be re-entered into the atmosphere in a controlled manner.

# X-38 Crew Return Vehicle

The X-38 was conceived as an emergency crew return vehicle for the ISS. The population of the ISS depends on how many human-rated craft are attached. This is currently 6, since there are two Soyuz capsules docked at any given time. When a new crew goes up, they take the capsule that has been docked to the station the longest, back down.

The X-38 had many innovative features. It could be

flown to a safe landing by controllers on the ground. A para-sail was considered for landing. It was a lifting-body architecture

A lot of the vehicle was designed with COTS components. The thermal tiles were similar to those used on the Shuttle. The flight computers came from commercial aircraft. The Navigation system was in use on Navy fighters. It would use GPS navigation.

Scaled Composites got a contract in 1996 to build three full scale airframes for testing. The first was delivered that year. Drop tests were conducted from a B-52 aircraft from 45,000 feet altitude. Near-trans-sonic speeds were achieved. A drogue parachute slowed this to 60 miles per hour. Flight control was autonomous, with a stand-by pilot/controller on the ground.

NASA also partnered with the European Space Agency and the German Space Agency on this project. In usage, the X-38 would be carried to the ISS onboard the Shuttle. One rocket plane as cargo for the other.

As well as things were going, cost-overruns caused the cancellation of the program in 2002. The three flight test models can be seen at the Strategic Air and Space Museum, Ashland Nebraska; outside of Building 49 at Johnson Space Center; and at Evergreen Aviation Museum in McMinnville, Oregon.

## Planned Follow-ons

This section discusses follow-on efforts to the current ISS. It is currently planned to decommission the facility. The Orbital Piloted Assembly and Experiment Complex (OPAEC) is a proposed Russian follow-on Space Station, using some modules from the ISS when it is decommissioned. Some of the station modules were not designed for disassembly, and a controlled reentry is about the only option at the current time.

# Deep Space Gateway

The Deep Space Gateway (DSG) is a NASA Project for a crewed station in cislunar space. It is intended as a jumping-off point. The Orion crewed vehicle is scheduled to be used for this effort. The Gateway would be located in a halo orbit around the Moon. By that, we mean that the spacecraft would be visible to Earth for its entire orbital path. Ion thrusters are proposed for station-keeping. These use electrical power for accelerating various (usually, inert) gasses to high velocity, rather than using fuel and oxidizer. The thrust is generally low, but can be continued for long periods of time.

The Power/Propulsion Module (PPM) will have large solar arrays for power, 40 kW is baselined. The Cislunar Habitation Module will join the PPM later in orbit, and will provide living and work space. The planned Gateway and Logistics module will join next, providing experiment space and supplies. An airlock module will be added later, to enable EVA operations.

Although we have data on 1 year duration mission close to Earth, the DSG will provide more information on long duration human missions in an environment away from our home planet. All of the ISS partners, US (Nasa), Russia (Roscosmos), Europe (ESA), Japan (JAXA), can Canada (CSA) are participating, and form the nucleus of further human exploration of the solar system.

The project presents daunting challenges in design, testing, delivery, logistics, and operations. The lunar vicinity provides a effective location for expeditions to other places in the solar system. And, it's been a while since we have stepped foot on the Moon. In the mean time, the discovery of water ice at the poles, from remote sensing, is most interesting for a lunar base.

The Gateway will serve, then, as a "enabling infrastructure" for further exploration. It is planned to be placed between the Earth and the Moon, in a null in the gravity field called a libration point. The Gateway will serve as the starting point and the mothership for lunar exploration. Telerobotic rovers on the Moon can be operated from the DSG with ease, although similar operation from Earth suffers from excessive communication time. The DSG offers return to Earth in a matter of days.

This will kick off with Expedition-1 of the Artemis Program, an uncrewed flight of NASA's new Space Launch System (SLS) Vehicle, and Orion Spacecraft.

This is scheduled, at the moment, for 2019. Various subsystems for the Orion capsule are now under test at the ISS. NASA estimates a rate of one SLS flight per year is possible, after the second flight. The basic SLS can lift 105 metric tons to Low Earth orbit, and the advanced vehicle will be capable of lifting crew and 10 metric tons to the lunar vicinity., or more than 40 tons, non-crewed.

The DSG is designed to use electric propulsion for station-keeping, eliminating the need for thruster fuel, and based on the amount of available sunlight. These will be ion thrusters, that use a monopropellant.

The DSG will result in a one-year crewed mission near the moon, to validate the concept of a flight to Mars. It is not so much the distance to Mars, as the relative orbital positions of the two planets in their solar orbits. The mechanics of the transfer orbit were worked out in 1925 by German scientist Walter Hohmann. In his 1928 book, *A Daring Trip to Mars,* Max Valier shows that one of the most efficient methods of reaching Mars from Earth involves a non-intuitive Venus fly-by. People have been thinking about this for a while.

The Russians are also eyeing a Lunar Orbital Station sometime after 2030.

# Deep Space Habitat

The Deep Space Habitat (DSH) was proposed in 2012, to

support human exploration beyond low Earth orbit, as a stepping stone to lunar, asteroid, and Mars missions. It will utilize on-orbit experience with the ISS, and the new Orion capsule. The goal is to have a crew living and working for up to 1 year. An ideal location for the Habitat would be the L1 cislunar Lagrange point, a null in the Earth-Moon gravity field. The advantage of staying at this point is that both the Earth, and the Moon are "downhill" in the gravity field. It is 150 million km from Earth, and 1.5 million KM from the Moon. Strangely, you can orbit a Lagrange point, even though there's nothing there. The Lagrange point on the opposite side of the Moon from the Earth is where the new James Webb Space Telescope is being placed. The Lagrange points, and there are 5 of them, are solutions in the restricted three body problem of orbital mechanics. The restriction is, one of the 3 body's much be much smaller than the other two. For any two bodies, there are 5 Lagrange points. Problem is, they are not the exact null points that we would like, but are perturbed by all the other bodies in the vicinity. You can trust me, or you can do the math.

The project is in Phase 2 of 3 as of this writing. Besides the Orion capsule for 4, the 60-day mission profile would utilize the Cyrogenic Propulsion System (Liquid hydrogen, liquid oxygen) assembly, a lab module, and an airlock. The MultiMission Space Exploration Vehicle may also be attached. This is a servicing craft for a crew of two, and, most importantly, it includes a toilet.

A 500-day mission is also baselined, requiring the

addition of a Multipurpose logistics module. The lessons learned from decades of various space station operations in orbit will be applied, as DSH will be further away from the home planet, and the cost of failures is much higher.

## Russian Lunar Orbital Station

The Russian Lunar Orbital Station is a 2007 proposed project for a lunar orbital station, and an eventual surface station. It will be based on MIR and ISS lessons learned. The project may see construction around 2030.

## Chinese Lunar Exploration Program

The Chinese see a lunar mission as the next logical step for them in space, as they gain technical and operational experience with their space station. They see commercial potential in Helium-3 mining, and exploitation of other lunar minerals. They have orbited the moon with unmanned craft, and deployed a rover on the surface. They have said that their next endeavor will be a sample return mission.

## LOP-G

The Lunar Orbital Platform - Gateway is a renaming and update to the Deep Space Gateway Program, basically a name change and some technical details updated. It will serve as staging point, in lunar orbit, for the Deep Space Transport, a re-usable crewed vehicle for Mars missions, using electrical and chemical propulsion.

The facility will have a power and propulsion element (PPE) derived from the DSG, with a mass of 8-9 metric tons, and be capable of supplying 50 KW of solar electric power for the ion thrusters. The project is in early concept phase, but some modules are being studied. These include a cis-lunar habitation module, compatible with the Orion capsule, a gateway logistics module, with a robotics arm. The Gateway airlock module will allow EVA activities, and could also be used as a short-term habitat. Lessons-learned from the ISS will be used for these modules, and certain technologies and assemblies are being tested on the station.

As of this writing, several NASA contracts have been let to Industry to implement the LOP-G. Maxar Technologies, of Colorado is to "develop and demonstrate power, propulsion, and communications capabilities." Maxar used to be Space Systems Loral. NASA selected the Space-X Falcon Heavy to launch the major components of the Gateway in late 2024.

# NASA Lunar Outpost

The NASA Lunar Outpost is an element of the Bush administration (2001-2009) *Vision for Space Exploration.* It was directed by Congress that the facility would be named the *Neil A. Armstrong Lunar Outpost.* The outpost location would be at one of the lunar poles. Since then, remote observations have revealed the presence of water ice in craters at the south Pole The South Poles remains in shadow, and sunlight does not reach the bottom of the craters. Besides the value of in situ water supplies, and the ability to produce hydrogen and oxygen from the water via solar-powered hydrolysis, the ice may contain records of the material of the early solar system. The water is a critical sustainable resource for a crewed lunar base. The Indian Chandrayaan lunar orbiter was responsible for this discovery.

One ideal structure for lunar bases (that the author has worked on) is lava tubes. These are found on the Earth as well, and the cooled lava provides a hard, sealed surface. It just needs to be capped with airlock doors, and no further exterior construction is required.

The outpost will consist of various modules for habitation and laboratory space, a solar array assembly, and a garage for a rover. There will also be a communications facility for the link to Earth. The facility was designed for a crew of four with 7-day visits during deployment, and up to 180 day operational missions.

## Commercial Efforts

Space can enable manufacturing facilities for new materials that can't be constructed in a gravity field. Other big moneymakers are orbiting vacation destinations, and in-space manufacturing. This topic will be discussed in detail in a separate book.

# Blue Origin
"...to build space hotels, amusement parks and colonies for 2 or 3 million people who would be in orbit. 'The

whole idea is to preserve the earth' he told the newspaper .... The goal was to be able to evacuate humans. The planet would become a park." Jeff Bezos, High School Class valedictorian, 1982, Miami Herald.

Blue Origin is a spaceflight services company founded by Jeff Bezos, of Amazon. The company "is developing technologies to enable private human access to space with the goal to dramatically lower costs and increase reliability." They are focusing on vertical take-off, vertical landing reusable vehicles. The name refers to Earth. Their spacecraft is named *New Shepard,* and has been flown un-crewed as of this writing. It went almost to space, reaching just shy of 100 km. The second flight did go beyond 100 km, and both the capsule and booster were recovered. They have reflown the hardware from the first mission, several times. Crewed flight will happen sometime in 2018. The company uses its own launch vehicles. The company is located in Washington State, and uses a launch facility in west Texas for sub-orbital launches. They have leased facilities at the Kennedy Space Center for orbital launches. Some technology from the earlier McDonald Douglas-NASA DC-X project influenced the Blue Origin.

The orbital rocket is named "*New Glenn*" after Astronaut John Glenn, The first stage will have seven of their BE-4 engines, and will be recoverable and reusable. The BE-4 engine production facility will be in Huntsville, AL. The rocket portion was recovered and reused a total of three times as of this writing.

The company has a NASA Space Act Agreement in place under the Commercial Crew Development Program. They are implementing their own spacestation space tourism and private research.

# Bigelow Aerospace Commercial Space Station

Bigelow Aerospace is an American space company, headquartered in Las Vegas, NV. It was founded in 1998, and focuses on inflatable modular units, carried to orbit. The company licensed the technology from NASA, and has Space Act agreements in place. The company has enhanced the technology.

Several modules have been sent to orbit, including Genesis-I (2006) and Genesis-II (2007). Both remain in orbit but are retired. Both have an expected life of 12 years in orbit, at which point they will reenter the atmosphere and burn. Both Genesis modules were heavily instrumented. Questions remain about their longevity and safety in the space environment.

The Bigelow Expandable Activity Module (BEAM) is in orbit, attached to the ISS. It was funded by NASA, and was launched on a SpaceX cargo mission. The emphasis of the inflatable modules is proving the radiation protection and debris shielding of the inflatable. It was launched in 2016, and will be evaluated through 2018.

Bigelow hopes to put one of their Expandable Bigelow Advanced Station Enhancement (XBASE) modules on the moon by 2021. This is referred to as the B330 Module. Space Complex Alpha is their next-generation commercial space station. Bigelow has launch agreements in place with SpaceX and Lockheed Martin, and has launched on the Russian Dneiper vehicle.

Bigelow himself came out of the hospitality industry, and hopes to set up space hotels and habitats to encourage space tourism. One of these was the 2005 concept, CSS (Commercial Space Station) *Skywalker*. He is also addressing manufacturing in space, and has said he has agreements in place with six nations to pursue this.

His other concepts include the Next-Generation Commercial Space Station, and Space Complex Alpha, consisting of dual Sundancer Modules and a BA330. The BA330 *Nautilus* module has an enclosed volume of 330 cubic meters. The Sundancer would have had a volume of 180 cubic meters, but the project was canceled. The constraining factor is the lack of crew transportation systems, consisting, at the moment, of the Soviet Soyuz. The U. S. has no current crew transportation for space, since the Shuttle fleet as been retired.

In March 2020, Bigelow suspended all activity's and laid off all staff, partially due to the Corona Virus.

# Orion Span

This company had a concept for a private commercial space station called Aurora in 2018. It would host 2 crew and up to 4 tourists. They shut down operations in 2021.

## Orbital Technologies Commercial Space Station

This project is a 2010 proposed commercial effort in orbit. Orbital Technologies is a Russian company. The facility would be compatible with the Soyuz and Progress crew delivery and logistics craft. The project has support from the Russian Federal Space Agency, but has no private funding as of this date.

## Axion Space

Axion, of Houston, envisions a commercial replacement for the ISS. This will be used for research, manufacturing, and space tourism as early as 2020. Before their station is ready, they plan to deliver tourists to the ISS, on a non-interference basis to the station's mission. Their modular station will be built in an incremental manner. As the ISS is decommissioned, the Axiom modules will separate, and go their own way. At this point, they will focus on tourism and in-space manufacturing as money-makers.

They are flying the first all-private mission to the ISS, currently slated for 2022

## SpaceX

Space-X has an impressive track record, including the first privately funded liquid rocket engine to orbit, the first to launch, recover, and reuse a rocket, the first private company to send a vehicle to the ISS, and the first reuse of an orbital rocket. Their Dragon capsule is designed in two versions. The cargo version has been in operation since 2010. The crewed version is scheduled to be tested in 2018 as of this writing. After the newer version is declared operational, the old version will still be used for cargo flights.

Their newer Dragon V2 capsule has a capacity of 7 crew. The Dragon capsules have an integral launch escape system, consisting of rocket engines. The vehicles are capable of autonomous rendezvous and docking. The onboard control is of the form of a tablet computer. They can parachute to a water landing, or do a dry landing, assisted by the external rocket motors. The capsule is partially re-usable, with a projected life of ten flights before refurbishment.

At the moment, the target launch price is $160 million. In a crewed version, this works out to $20 million per seat. Russia currently charges $76 mission per seat on Soyuz. All personnel will wear space suits while in the capsule.

## *Soviet Union/Russia*

This section presents the Soviet Union/Russia'a efforts in crewed space stations. Russia is a participant in the International Space Station.

# Salyut

Salyut ("salute") was the Soviet Union's first space station project. The program ran from 1971 to 1991, and involved 9 units, six of which were successfully manned. Salyut-1 was the first crewed Space Station. It used a series of onboard computers that were designed and produced by the Scientific Research Institute of Micro-Instruments near Moscow. There were two models with different missions, the Durable Orbital Station (DOS) was civilian, and the Almaz-OPS (Orbital piloted station) was military.

Salyut-1 was the first space station in orbit, launched two years before Skylab. A total of 1,697 occupied days were compiled.

On the first crewed flight to Salyut-1, the crew was unable to dock, due to a failure in the mechanism on their craft. A second mission was successful, and the crew occupied the facility for more than 3 weeks. Unfortunately, the crew as killed while returning, due to a faulty pressurization valve. They were not wearing pressure suits, as the cabin space did not allow that. Salyut 1 was intentionally reentered and destroyed after 175 days in orbit. The replacement unit did not reach orbit, due to a launch vehicle fault. A new mission, launched a few days before Skylab, had a failure that caused the orbit control thrusters to fire continuously, depleting the fuel supply. Ity reentered the atmosphere and burned.

The evolutionary architecture led to a second docking

port being added. The next to last unit was abandoned and reentered due to mold problems onboard. Salyut-7 went to orbit in 1983, and was operational for more than 8 years. It was visited by ten crewed spacecraft, six with long duration crews. The Salyut series led to Mir and the ISS. The heritage of modular design was proven, and continued.

## Almaz

Almaz ("Diamond") was a Soviet Military Reconnaissance station, identified as Salyut 2, 3, and 5. The program ran from 1973-1976. These Soyuz craft, were referred to as Orbital Pilot Stations (OPS). This program was similar to the USAF's MOL Project. The Almaz carried a 23mm automatic cannon, for self-defense. It was tested in orbit against a satellite target.

A private company, Excalibur Almaz, bought two partially completed but unflown Almaz hulls, as well as several capsules. The company estimates they saved $2 billion dollars in development costs. Theses they plan to use for Space Tourism, and possibly for manufacturing in space. The space cannon were probably not included in the deal. The company is headquartered on the Isle of Man and successfully participated in NASA's Commercial Crew Development Program. They also own four reusable reentry vehicles (RRV) from the Russian program, modeled on the Soyuz craft.

## MIR

MIR ("Peace") was a Soviet modular space station, intended for long duration, Earth-orbital mission. It was launched in 1986, had visits from 28 crews, and included other nationalities. It was de-orbited in 2001. It logged the longest duration space mission to that point, more than 437 days in orbit, by two cosmonauts.

The MIR space station used a series of computers, the Argon-16, the Salyut-4, and the Salyut-5. The Argon-16 was a 16-bit machine, with 6 kilobytes of random access memory, and 48 kilobytes of read-only memory. Sixteen interrupts are supported. It accommodates a wide variety of analog and digital Input/Output devices. It weighs 70 kilograms, and used 280 watts of power. It entered production in 1974. The Argon-12S flight computer was delivered to the Mir, and took over attitude and orbit control of the 240 ton station. It was in orbit for 3,644 days before reentering the atmosphere.

MIR was designed for a crew of 3, with 6 possible for short periods. The Space Shuttle Atlantis delivered the docking module for MIR. It had seven pressurized modules, and large solar arrays for power. It was assembled in orbit, from various modules, launched separately. There were a total of seven modules, and cargo cranes to assist in maneuvering exterior equipment.

After the basic station was in-orbit, 2 cosmonauts visited in 1986, and powered the station up. On the way back, they visited another orbital outpost, Salyut-7. Later, when the station was occupied, a Kvant module was having trouble docking. An EVA revealed a trash bag in the way. Seems you can't just leave the trash outside the door, in orbit, and have it picked up.

The station used 28 volt dc power, with taps at various locations. The Nickel-cadmium batteries were charged by the large solar panels. Initially, the station had 9 kilowatts available, later expanded to more than 16 kilowatts. MIR had two toilets, which fed into a water recovery system. Collected solid wastes were shipped back to Earth on a resupply vehicle. The facility also had a shower.

MIR was in a near-circular orbit of 354-375 kilometers. Atmospheric drag continually lowered the orbit, and it was reboosted by the resupply vehicles. The station had thrusters for attitude control and pointing. The station maintained an Earth-sea level atmospheric pressure and composition.

There were twenty-eight crews that occupied MIR for various times, each identified as an "Expedition." These varied in duration from 72 to 437 days. An expedition was 2-3 crew. Visitors were on the station for about a week. In total, 104 persons from 12 nations visited the orbital facility.

There were two non-cosmonauts visiting Mir. These were Helen Sharman, the first Briton to fly in space, and Japanese TV reporter Toyohiro Akiyama in 1990, as a business trip. Cosmonauts from at least 5 countries visited MIR, as did U.S. Astronaust Jerry Linneger and Shannon Lucid.

During a docking system test in 1977, the docking vehicle collided with the Station's solar array, bounced into one of the modules, and punctured it. This lead to an emergency depressurization of the station. The crew responded quickly and contained the damage. No one was injured. In another incident, in 1997, there was an onboard fire caused by some of the oxygen generating equipment. This filled the facility with toxic fumes. Quick reaction by the crew resulting in containing the damage, and no injuries were reported. At the time, U. S. Astronaut Linenger was aboard. This has been the most severe fire about an orbiting space station so far. Again, lessons were learned that applied to the planned ISS. Close-out of Mir began in 1998

A resupply craft collided with the solar arrays during a test of the manual docking system. The module was holed, and was leaking the onboard atmosphere to space. Again, quick action on the part of the crew saved the day when they isolated to module by means of an airlock. Unfortunately, access to the module was lost. Later, the onboard crew was able to patch the leak, and open the airlock again.

In the 1990s samples of extremophile molds were taken at Mir. Ninety species of micro-organisms were found , four years after the station's launch. By the time of its decommission in 2001, the number of known different micro-organisms had grown to 140. As space stations get older, the problems with contamination get worse. Molds that develop aboard space stations can produce acids that degrade metal, glass and rubber. The molds in Mir were found growing behind panels and inside air-conditioning equipment. The molds also caused a bad odor, which was often cited as visitors' strongest impressions.

Some biologists were concerned about the mutant fungi being a major microbiological hazard for humans, and reaching Earth in the splashdown, after having been in an isolated environment for 15 years. Similar molds are found in the ISS.

In 1999, crews arrived to begin decommissioning the station, as the ISS project was ramping up. The station reentered and burned, landing in the South Pacific in March of 2001. The facility had been designed with a 5-year lifetime goal, but lasted 15.

MIR-II refers to a follow-on project, which eventually got incorporated into the ISS. The module used was the Zvexda ("star"). It is the basis for the current station's life support system.

## Shuttle-Mir Program

The Shuttle-Mir program was a demonstration of cooperation in Space, and led to the combination of two space station programs of the United States and Russia into one facility, the International Space Station. The U.S. wanted to take advantage of the Russian on-orbit experience, and Russia needed hard cash. This was a successful joint endeavor, continuing to this writing, regardless of the motivations.

Shuttle mission STS-63 flew to the MIR station in 1995. There was no attempt to dock; only a fly-around. There would be a total of nine Shuttle flight to Mir, delivering new modules for station expansion, crew exchange, and logistics supplies. American astronauts logged close to a thousand days onboard MIR. The station was a follow-on to the earlier Salyut Stations. It was larger and more comfortable, and took advantage of lessons-learned in orbit. With the docked Shuttle, the facility massed 250 metric tons, the world's largest facility in space.

The MIR Station was de-orbited successfully in March of 2001. It had flow three times its projected life.

# Opsek

The Orbital Piloted Assembly and Experiment Complex is a Russian proposed modular space station in low Earth orbit. This $3^{rd}$ generation system would consist of some of the Russian segments on the existing ISS, as that facility is decommissioned. The new facility would be targeted to research, and orbital assembly of large modules. If this path is taken, this will represent the $12^{th}$ Russian Space Station.

Some of the Russian building block modules from the ISS will be removed from that structure, and built into the more compact OPSEK. Theses would include the Poisk Mini-Research Module-2 (MRM-2) and Nauka, the multipurpose laboratory module, which has six docking ports. The latter is the main Russian Laboratory. The Node module would also be reused. It has docking ports for Soyuz and Progress-M. The node module will become the core of the new station. A new Science-Power module will be included for Opsek.

# Chinese Tiangong

In Chinese, the name Tiangong means *Heavenly Palace,* quite appropriate for a space station. The goal is to launch a third generation space station, similar to the Russian MIR, by 2020. Tiangong-1 was a 8-ton space laboratory, launched in 2011. The follow-on Tiangong-2 was launched in the Fall of 2016. Tiangong-3 was around 20 tons, and included a cargo resupply vessel. It supports an onboard crew of three, and there were 2 sets of crew that spent time in space. The life of the facility has been extended to beyond 2020, but will not receive any further crews.

China uses the Jiuquan Satellite Launch Center in the Gobi desert of Inner Mongolia. It is one of three Chinese launch facilities, and is suited for large orbital inclination angles. The launch of China's first satellite, and first crewed mission were from this facility.

Tiangong-1, sometimes referred to as a docking target, consists of a propulsion (resource) module and a pressurized module, with a docking mechanism at either end. The docking port of the experiment section supports automated docking. Its length is 10.5 meters, the diameter is 3.4 meters, and it has a mass of 8,000 kilograms. It was launched in 2011, and is intended for short stays by a crew of three.

The Tiangong-2 space laboratory was launched in 2016. It is a conglomerate of two previous space station projects. The Station is supplied by the Tianzhou capsule, with a capacity of 6,500 kg. The station was reentered into the atmosphere in July of 2019, and fell into a deep section of the Pacific.

Following in the footsteps of MIR and the ISS, the large Chinese Station, Tiangong-3, is scheduled to be assembled in orbit in 2020. It will consist of a Cabin module, a Laboratory Module, a Shenzhou crew vehicle, and a cargo resupply ship Tianzhou, the "Heavenly Vessel." The later is based on the Tiangong-1, and was launched by the Long March-7 rocket. It is designed to last for 10 years, and will mass around 60,000 kg. It is designed for three full-time residents.

There was to be a Tiangong-3, but it was replaced by a new modular design.

The Chinese Large Modular Space station, is about 1/6$^{th}$ of the ISS in size and weight. The core module Tianhe ("Harmony of Heaven") was launched in 2021. It is more than 18 meters long, and had a launch weight of 22 tons. The lab/cabin module is more than 14 meters long. The crewed module is Shenzhou. There is also a cargo craft, based on Tiangong. The station has a planned lifetime of 10 years.

## India

ISRO, the Indian Space Research Organization, is considering a space station project, possible in conjunction with other nations., in the next few years.

## *Afterthought*

The author hopes that by the time a revised edition of this book comes out, he will be able to speak of at least a lunar habitat, if not one on Mars. Actually, he hopes he can write said book from that facility.

# Bibliography

Baker, David *The History of Manned Space Flight*, 1982, Crown Publishers, ISBN-051754377X.

Baker, David, *International Space Station: An Insight into the History, development, collaboration, production, and role of the Permanently Manned Earth-orbiting Complexes*, 2016, ISBN-978-0857338396.

Bentley, Matthew A. *Spaceplanes: From Airport to Spaceport*, 2009, Springer, ASIN-B008BB7HQA.

Belfiore, Michael *Rocketeers, How a Visionary Band of Business Leaders, Engineers, and Pilots is Boldly Privatizing Space*, Harper Collins, 2008, ISBN-0061149039.

Bergaust, Erik, *Werner von Braun*, National Space Institute, 1976, ISBN-0917680014.

Bluth, B. J. Helpple, Martha Soviet Space Stations as Analogs, 2nd ed, August 1986, (updated with MIR 1987). produced under NASA Grant NAGW-659, avail: https://ntrs.nasa.gov/archive/nasa/casi.ntrs.nasa.gov/19870012563.pdf

Buckley, James *Home Address: ISS: International Space Station,* Smithsonian, Sep 1, 2015.

Burrough, Bryan *Dragonfly, An Epic Adventure of*

*Survival in Outer Space,* 1988, Harper Collins, ISBN-0-88730-783-3.

Buss, Jared S. *Willy Ley: Prophet of the Space Age*, 2017, University Press of Florida, ISBN-0813054435.

Chladek, Jay *Outposts on the Frontier: A Fifty-Year History of Space Stations (Outward Odyssey: A People's History of Spaceflight*, 2017 U. Nebraska Press, ISBN-0803222920.

Choi, Charles Q. "In Race for Private Space Stations, It's U.S. Versus Russia," Nov. 12, 2010. Space.com, avail: https://www.space.com/9518-race-private-space-stations-russia.html

Compton, William David *Living and Working in Space: The NASA History of Skylab*, Dover, 2011, ISBN-0486482189.

Congress of the United States, Office of Technology Assessment, *Salyut: Soviet Steps Toward Permanent Human Presence in Space, A Technical Memorandum*, 1983, ASIN-B0065RY5FI.

Eckhart, Peter *The Lunar Base Handbook,* 1999, 1st ed, McGraw-Hill Primis Custom Publishing, ASIN-B01A1MSBRK.

Ellsworth-Smith, Lincoln *A Day in the Life Aboard the International Space Station*, 2015, ASIN B00V3RNAXS.

Evans, Ben *Partnership in Space: The Mid to Late Nineties,* Springer Praxis, 2013, ISBN-1461432774.

Ezell, Edward Clinton and Ezell, Linda Neuman *The Partnership, a History of the Apollo-Soyuz Test Projec*t, (NASA SP-4209), The NASA History Series, 1978, ASIN-B0000E96NI.

Froehlich, Walter *Apollo Soyuz,* 1976, NASA, EP-109, ASIN-B0006CWXQQ.

Gibson, Karen Bush *Women in Space: 23 Stories of First Flights, Scientific Missions, and Gravity-Breaking Adventures (Women of Action),* 2014, Chicago Review Press, ASIN-B00HXZN0KW.

Goodwin, Robert *Rocket And Space Corporation Energia,* ISBN-1896522815.

Griffin, Brand "Skylab II: Making a Deep Space Habitat from a Space Launch System Propellant Tank, "March 27, 2013 Future In-Space Operations Colloquium, Future In-Space Operations Working Group. Avail: http://spirit.as.utexas.edu/~fiso/telecon13-15/Griffin_3-27-13/.

Hale, Edward Everett *The Brick Moon,* 1899, available in 2017 reprint, ISBN-1374824542.

Harland, David M *Creating the International Space Station,* Springer-Praxis, 2002, ISBN-1852332026.

Hitt, David *Homesteading Space: The Skylab Story*, 2008, University of Nebraska Press, ASIN-B003NHSBNU

Holt, Nathalia *Rise of the Rocket Girls: The Women Who Propelled Us, from Missiles to the Moon to Mars*, 2016, Hachette Book Group, ASIN-B013CATQPY.

Ivanovich, Ujica *Salyut - The First Space Station: Triumph and Tragedy* (Springer Praxis Books), 2008, ISBN-0387735852

Jahns, Richard K. *Skylab Explores the Earth*, 1977, NASA-SP380, ASIN-B0000EH017.

Launius, Roger D. *Space Stations: Base Camps to the Stars*, 2003, Smithsonian Books, ISBN-1588341208.

Ley, Willy and Aldrin, Buzz, *Rockets, Missiles, and Men in Space: A Definitive Account of the History of Space*, 1966, Viking Press, ASIN-B004H3G0WI.

Ley, Willy *Rockets, Missiles & Space Travel*, 1958, Viking Press, reprinted Signet, 1969, ASIN-B001Q9KTGE.

Ley, Willy *Missiles, Moonprobes, and Megaparsecs*, 1964, Signet, ASIN- B000HWU3MG.

Ley, Willy *Space Stations: Adventures in Space*, 1958,

Guild Press, ASIN-B000LB5OMC.

Ley, Willy, Rockets, *The Future of Travel Beyond the Stratosphere*, 1945, Viking Press, ASIN- 0007E7IC2.

Linenger, Jerry *Off the Planet: Surviving Five Perilous Months Aboard the Space Station Mir*, 2000, McGraw-Hill, ISBN-007136112X.

Linenger, Jerry *Letters from MIR: An Astronaut's Letters to His Son*, 2002, McGraw-Hill, ISBN-0071400095.

NASA, *Inside the International Space Station (ISS): NASA International Space Station Familiarization Astronaut Training Manual - Comprehensive Review of ISS Systems*, 2011, ASIN-B006O403MG.

NASA, *Reference Guide to the International Space Station*, 2014, ASIN-B00M3K6LP8.

NASA, *Mir Space Station NASA Astronaut Training Manual - Complete Details of Russian Station Onboard Systems, History, Operations Profile, EVA System, Payloads, Progress, Soyuz, Salyut*, 2012, ASIN-B008RIC140.

NASA, *Shuttle-Mir: The United States and Russia Share History's Highest Stage,* NASA-SP-2001-4225, 2011, ASIN-B006A8G5OG.

Newell, Peter von Polentz, Woldgang *Die Rakete / The*

*Rocket Book,* (in German), 2017, Amalienpresse, ISBN-3939904147.

Nixon, David *International Space Station: Architecture beyond Earth*, 2017, Circa Press, ISBN 0993072135.

Noordung, Hermann, Potocnik, Herman *The Problem of Space Travel: The Rocket Motor,* 2015, ASIN-B015EYQR9O.

Oberth, Herman *Die Rakete Zu Den Planetenräumen*, reprint,2013, in German, ISBN-348674187X.

Portree, David S. F, *Mir Hardware Heritage*, March, 1995, NASA RP 1357.

Powell-Willhite, Irene E. *The Voice of Dr. Wernher von Braun: An Anthology*, 2007, Collectors Guide Publishing, ISBN-1894959647.

Seedhouse, Erik *Bigelow Aerospace: Colonizing Space One Module at a Time*, 2015, Springer, ISBN-3319051962.

Seedhouse, Erik *SpaceX's Dragon: America's Next Generation Spacecraft,* 2016, Springer, ISBN-3319215140.

Seedhouse, Erik *SpaceX: Making Commercial Spaceflight a Reality,* 2013, Praxis, ISBN-1461455138.

Seedhouse, Erik *Virgin Galactic: The First Ten Years*, Springer, 2015, ISBN-3319092618.

Seedhouse, Erik *Lunar Outpost: The Challenges of Establishing a Human Settlement on the Moon,* 2008, Springer, ISBN-0387097465.

Shayler, David J. Assembling *and Supplying the ISS: The Space Shuttle Fulfills Its Mission* (Springer Praxis Books), 2017, Springer, ISBN-3319404415.

Shayler, David J. *Linking the Space Shuttle and Space Stations: Early Docking Technologies from Concept to Implementation,* 2017, Springer, ISBN-3319497685 .

Sherr, Lynn *Sally Ride: America's First Woman in Space,* 2014, Simon & Schuster, ASIN-B00GEEB99W.

Simpson, Theodore R. *The Space Station, An Idea Whose Time has Come,* IEEE, 1985, ISBN-0879421827.

Smitherman, David *Habitat Concepts for Deep Space Exploration,* 2014, NASA, ASIN-B01ED7JF10.

Steven-Boniecki, Dwight *Skylab 1 & 2*, 2015, Apogee Books, ISBN-1926592271.

Steven-Boniecki, Dwight *Skylab 3, The NASA Mission Reports,* 2016 Apogee Books, ISBN-192659228X .

Tsiolkovsky, Konstantin E. *Selected Works of Konstantin*

*E. Tsiolkovsky*, 2004, University Press of the Pacific, ISBN-141021825

U. S. Government, DoD, USAF, Center for the Study of National Reconnaissance, *The Dorian Files Revealed: A Compendium of the NRO's Manned Orbiting Laboratory (NRO) Documents, Photoreconnaissance, Spy in the Sky, Blue Gemini, Air Force Space Station, Dyna-Soar, Apollo Study*, 2017, ISBN-1521163073.

Vance, Ashlee *Elon Musk: Tesla, SpaceX, and the Quest for a Fantastic Future*, 2015, Ecco, ISBN-0062301233.

World Spaceflight News & NASA *Inside the International Space Station (ISS): NASA International Space Station Familiarization Astronaut Training Manual - Comprehensive Review of ISS Systems*, 2011, ASIN-B006O403MG.

## Resources

- www.nasa.gov
- https://history.nasa.gov/
- https://history.nasa.gov/tindex.html#5
- http://klabs.org/history
- http://www.airspacemag.com/space/future-construction-space-180956237/
- Encyclopedia Astronautica, http://www.astronautix.com/
- bigelowaerospace.com
- http://www.indianspacestation.com/

- Wikisource: Mir Hardware Heritage/Part 2 - Almaz, Salyut, and Mir#2.1.6 Shuttle-Salyut . 281973-1978.3B 1980s.29
- Andy Thomas (U.S. Astronaut on MIR), https://history.nasa.gov/SP-4225/documentation/thomas-letters/letters.htm
- Vectors website - http://vc.airvectors.net/idx_sci.html
- https://www.nasa.gov/mission_pages/station/main/suni_iss_tour.html
- Mutant space microbes attack ISS; avail: https://www.rt.com/news/iss-bacteria-mir-mutation-765/
- Wikipedia, various.

# Glossary of Terms

ABMA – Army Ballistic Missile Agency, Redstone Arsenal, Huntsville, Alabama.
AGC – Apollo Guidance Computer
AIAA – American Institute of Aeronautics and Astronautics.
ALU – arithmetic logic unit
AMD – Aircraft Missiles Division, Fairchild Hiller, Hagerstown, MD.
AOMC – Army Ordnance Missile Command – 1958, Redstone Arsenal, JPL, WSPG.
Apogee – farthest point in the orbit from the Earth.
ARPA – Advanced Research Projects Agency.
ASC – Agence Spatiale Canadienne, see also CSA.
ASIN – Amazon Standard Inventory Number
ASTP – Apollo-Soyuz Test Program.
Astrionics – electronics for space flight.
ATOLL - Acceptance Test or Launch Language.
Baikonur Cosmodrone – Russian launch site, in Kazakstan.
BEAM - Bigelow Expandable Activity Module.
BEO – beyond Earth orbit.
Bifacial – two sided.
BGA – ISS ball gimbel assemblies
Blooster – balloon-based launch vehicle.
BP – boilerplate. Mechanical model.
CASIS – Center Center for Advancement of Science in Space, non-profit.
CBM – common berthing mechanism
CCICap – NASA's Commercial Crew Integrated

Capability program.
CLEP – Chinese Lunar Exploration Program.
CMG – control moment gyro.
CNES – Centre national d'etudes spatiales, French space agency.
COTS – Commercial, off the shelf
Cps – cycles per second; cyrogenic propulsion stage.
Cpu – central processing unit.
CSA – Canadian Space Agency, see also ASC.
CSS – China Space Station.
CST – FAA Office of Commercial Space Transportation.
CT – communications and tracking
Cygnus – Orbital-ATK automated cargo vehicle for ISS
Cyrogenic – pertaining to very low temperatures.
DAM – damage avoidance maneuver.
DCM – (ISS) docking cargo module.
DoD – (U. S.) Department of Defense.
Downmass – stuff coming back to Earth.
DSG – Deep Space Gateway.
DSH – deep space habitat.
DSL – digital subscriber link, a bandwidth limited networking architecture, mostly obsolete.
DTM – dynamic test model, for structural tests.
ECLSS – Environmental Control & Life Support system.
ELC – Express Logistics carrier (ISS)
ELM-ES – (Japanses) Experiment Logistics Module, external section.
ELM-PS – (Japanese) Experiment Logistics Module, Pressurized Section
Ephemeris – position information data set for orbiting bodies, 6 parameters plus time.

EPS – electrical power system
ESA – European Space Agency.
ESP – External Storage Platform (ISS)
ETCS – ISS external thermal control system
EuTEF – European technology exposure facility – a platform outside the Columbus module on the ISS.
EVA – extra-vehicular activity – going outside in a space suit.
FAA – (U. S.) Federal Aviation Administration.
FPMU – ISS floating point measurement unit
Gimbal – pivoted support, allowing rotation about 1 axis.
GNC – Guidance, Navigation, and Control
Gpm – gallons per minute.
GSFC – NASA Goddard Space Flight Center, Greenbelt, MD.
Gyro – device to measure angular rate.
H1 – Rocketdyne engine, used on Saturn-I first stage.
HTV – H-II Transfer Vehicle, Japanese for ISS.
IBM – International Business Machines Company.
ICS – Interorbital Communications System, Japanese data link between ISS and Earth.
IDSS – International Docking System Standard.
IGA - (ISS) InterGovernmental Agreement
IRBM – Intermediate Range Ballistic Missile.
ISBN – international standard book number.
ITS – Integrated Truss Structure of the ISS.
ISP – specific impulse. Measure of efficiency of rocket engine. Units of seconds.
ISRO – Indian Space Research Organization
JAXA – Japan Aerospace Exploration Agency.
JEM – Japanese Experiment Module, ISS.

JEMRMS – Japanese Experiment Module Remote Manipulator System.
JPL – Jet Propulsion Laboratory, Pasadena, CA.
JSC – Johnson Space Center, Houston, Texas.
Jupiter – ICBM, 3-stage. Developed by von Braun Team.
Kibo – Japanese module on ISS
Kvant – Russian, Quantum.
KSC – NASA Kennedy Space Center, launch site, Florida.
KURS – Russian automated docking system for Soyuz.
L2 – second of 5 Lagrange points, a null in the gravity field in the restricted 3-body problem.
LAS – launch abort system
Lbf – pounds, force.
LC-37 – Launch Complex – 37 at KSC.
LEO – low Earth orbit.
LES – Launch Escape System.
LH2 – liquid hydrogen.
LOS – lunar orbital station; loss of signal.
LOX – liquid oxygen, boils at -297 F.
LVDA – Launch Vehicle Data Adapter.
MBS – ISS Mobile Base system
MDM – multiplexer/demultiplexer,
MET – mission elapsed time.
Mev – million electron volts, measure of energy of a particle.
MINITRACK – "Minimum Trackable Satellite " U. S. satellite tracking network, 1957.
MIT – Massachusetts Institute of Technology.
MLM – (Russian) Multipurpose Laboratory Module, part of ISS.

MMOD – ISS Micro Meteroid orbital debris
MMSEV – MultiMission Space Exploration Vehicle,
MOU – memorandum of understanding.
MPLM – Multi-purpose Logistics Module.
MRM – Russian Mini Research Module, part of ISS.
MSC – Manned Space Center, Houston, TX. Renamed JSC.
MSFC – NASA Marshall Space Flight Center, Huntsville, AL.
MSS – Mobile servicing system (ISS).
m/s – meters per second.
MY – ISS mobile transporter
Nadir – the point directly below.
NASA – National Aeronautics and Space Administration.
NASCOM – NASA Communications Network. Worldwide, operated by GSFC.
NextSTEP – Next Space Technologies for Exploration Partnerships, NASA Program.
Node module – (Russian) part of the ISS, docking module, allowing access for Soyuz and other craft.
NORAD – North American Air Defense.
NRL – Naval Research Lab, Washington, DC.
NRO – (U. S.) National Reconnaissance Office.
NTIS – National Technical Information Service (www.ntis.gov).
OAMS – Orbit Attitude and Maneuvering System
Ogive – a pointed arch shape.
Opsek – (Russian) Orbital Piloted Assembly and Experiment Complex.
Orlan – Russian space suite, used on ISS.
ORU – Orbital Replacement Unit

PCU – (ISS) plasma contactor unit – anti-static devices for ISS.
Perigee – closest point in the orbit from the Earth.
PGNCS – Primary Guidance, Navigation, and Control system for Apollo.
PGSC – Portable General Support computer – ISS laptop.
PGT – ISS Pistol grip tool
PMA – ISS Pressurized mating adapter.
PMCU – Power Management Control Unit.
POGO – longitudinal oscillation in liquid-fueled rocket motors that can lead to failure.
Poisk – (Russian) airlock on the ISS.
Pregnant Guppy – large cargo aircraft operated by Aero Spacelines 1963-1979.
PTCS – ISS Passive thermal control system
PVCU – ISS Photo Voltaic Control Unit.
R&D – research & development.
Redstone – Army missile developed by the von Braun team. Used for Mercury manned flights.
Redstone Arsenal – Army R&D facility in Huntsville, AL. Later became NASA MSFC.
RGA – ISS rate gyro assembly
RKA – Roscosmos State Corporation, (Russia).
RMS – remote manipulator system, Shuttles "arm"
ROS – Russian orbital segment of the ISS.
RP-1 – rocket propellant-one, highly refined kerosene.
SA – x – Saturn-Apollo – flight x.
SAO – Smithsonian Astrophysical Observatory.
SARJ – ISS Solar Alpha Rotary Joint
SEPS – ISS Secondary electrical power system.
SFOG – Russian Solid Fuel Oxygen generation system.

SI – System International – the metric system.
S-IC – first stage of the Saturn V
S-II – second stage of the Saturn V
S-IVB – third stage of the Saturn V
S-IV – second stage of Saturn 1 rocket.
SLS – Space Launch System.
S&M – ISS structures and mechanisms
SPDM – (ISS) Special Purpose Dexterous Manipulator - "Dextre" gripping attachment or hand.
SSRMS – Space Station Remote Manipulator System.
STADAN – Space Tracking and Data Acquisition Network.
TCS – thermal control system
Titan – ICBM and NASA/USAF launch vehicle.
TM – Technical Manual.
TRRJ – ISS Thernal Radiator Rotary Joint
TTL – transistor-transistor logic.
TUS – ISS trailing umbilical system
UDM – ISS universal docking module.
Ullage – residual fuel or oxidizer in a tank after engine burn is complete.
USAF – United States Air Force.
USOS – U. S. Orbital Segment, ISS.
V-2 – German World War-II missile developed by the von Braun Team.
VfR – (German) Verein für Raumschiffahrt, amateur rocket society, pre-World War-II.
WETA – ISS wireless external transceiver assembly
WSMR – White Sands Missile Range, New Mexico.
XBASE - Expandable Bigelow Advanced Station Enhancement.

Zenith – the point directly above.
Zombie-Sat – dead satellite, in orbit
Zvezda – Russian ISS Module.

# If you enjoyed this book, you might also be interested in some of these.

Stakem, Patrick H. *16-bit Microprocessors, History and Architecture*, 2013 PRRB Publishing, ISBN-1520210922.

Stakem, Patrick H. *4- and 8-bit Microprocessors, Architecture and History*, 2013, PRRB Publishing, ISBN-152021572X,

Stakem, Patrick H. *Apollo's Computers*, 2014, PRRB Publishing, ISBN-1520215800.

Stakem, Patrick H. *The Architecture and Applications of the ARM Microprocessors*, 2013, PRRB Publishing, ISBN-1520215843.

Stakem, Patrick H. *Earth Rovers: for Exploration and Environmental Monitoring*, 2014, PRRB Publishing, ISBN-152021586X.

Stakem, Patrick H. *Embedded Computer Systems, Volume 1, Introduction and Architecture*, 2013, PRRB Publishing, ISBN-1520215959.

Stakem, Patrick H. *The History of Spacecraft Computers from the V-2 to the Space Station*, 2013, PRRB Publishing, ISBN-1520216181.

Stakem, Patrick H. *Floating Point Computation*, 2013, PRRB Publishing, ISBN-152021619X.

Stakem, Patrick H. *Architecture of Massively Parallel Microprocessor Systems*, 2011, PRRB Publishing, ISBN-1520250061.

Stakem, Patrick H. *Multicore Computer Architecture,* 2014, PRRB Publishing, ISBN-1520241372.

Stakem, Patrick H. *Personal Robots*, 2014, PRRB Publishing, ISBN-1520216254.

Stakem, Patrick H. *RISC Microprocessors, History and Overview,* 2013, PRRB Publishing, ISBN-1520216289.

Stakem, Patrick H. *Robots and Telerobots in Space Applications*, 2011, PRRB Publishing, ISBN-1520210361.

Stakem, Patrick H. *The Saturn Rocket and the Pegasus Missions, 1965,* 2013, PRRB Publishing, ISBN-1520209916.

Stakem, Patrick H. *Visiting the NASA Centers, and Locations of Historic Rockets & Spacecraft,* 2017, PRRB Publishing, ISBN-1549651205.

Stakem, Patrick H. *Microprocessors in Space*, 2011, PRRB Publishing, ISBN-1520216343.

Stakem, Patrick H. Computer *Virtualization and the Cloud*, 2013, PRRB Publishing, ISBN-152021636X.

Stakem, Patrick H. *What's the Worst That Could Happen? Bad Assumptions, Ignorance, Failures and Screw-ups in Engineering Projects*, 2014, PRRB Publishing, ISBN-1520207166.

Stakem, Patrick H. *Computer Architecture & Programming of the Intel x86 Family*, 2013, PRRB Publishing, ISBN-1520263724.

Stakem, Patrick H. *The Hardware and Software Architecture of the Transputer*, 2011, PRRB Publishing, ISBN-152020681X.

Stakem, Patrick H. *Mainframes, Computing on Big Iron*, 2015, PRRB Publishing, ISBN- 1520216459.

Stakem, Patrick H. *Spacecraft Control Centers*, 2015, PRRB Publishing, ISBN-1520200617.

Stakem, Patrick H. *Embedded in Space*, 2015, PRRB Publishing, ISBN-1520215916.

Stakem, Patrick H. *A Practitioner's Guide to RISC Microprocessor Architecture*, Wiley-Interscience, 1996, ISBN-0471130184.

Stakem, Patrick H. *Cubesat Engineering*, PRRB Publishing, 2017, ISBN-1520754019.

Stakem, Patrick H. *Cubesat Operations*, PRRB Publishing, 2017, ISBN-152076717X.

Stakem, Patrick H. *Interplanetary Cubesats*, PRRB Publishing, 2017, ISBN-1520766173 .

*Stakem, Patrick H. Cubesat Constellations, Clusters, and Swarms, Stakem,* PRRB Publishing, 2017, ISBN-1520767544.

Stakem, Patrick H. *Graphics Processing Units, an overview*, 2017, PRRB Publishing, ISBN-1520879695.

Stakem, Patrick H. *Intel Embedded and the Arduino-101, 2017,* PRRB Publishing, ISBN-1520879296.

Stakem, Patrick H. *Orbital Debris, the problem and the mitigation*, 2018, PRRB Publishing, ISBN-*1980466483*.

Stakem, Patrick H. *Manufacturing in Space*, 2018, PRRB Publishing, ISBN-1977076041.

Stakem, Patrick H. *NASA's Ships and Planes*, 2018, PRRB Publishing, ISBN-1977076823.

Stakem, Patrick H. *Space Tourism*, 2018, PRRB Publishing, ISBN-1977073506.

Stakem, Patrick H. *STEM – Data Storage and Communications*, 2018, PRRB Publishing, ISBN-

1977073115.

Stakem, Patrick H. *In-Space Robotic Repair and Servicing*, 2018, PRRB Publishing, ISBN-1980478236.

Stakem, Patrick H. *Introducing Weather in the pre-K to 12 Curricula, A Resource Guide for Educators*, 2017, PRRB Publishing, ISBN-1980638241.

Stakem, Patrick H. *Introducing Astronomy in the pre-K to 12 Curricula, A Resource Guide for Educators*, 2017, PRRB Publishing, ISBN-198104065X.
Also available in a Brazilian Portuguese edition, ISBN-1983106127.

Stakem, Patrick H. *Deep Space Gateways, the Moon and Beyond*, 2017, PRRB Publishing, ISBN-1973465701.

Stakem, Patrick H. *Exploration of the Gas Giants, Space Missions to Jupiter, Saturn, Uranus, and Neptune*, PRRB Publishing, 2018, ISBN-9781717814500.

Stakem, Patrick H. *Crewed Spacecraft*, 2017, PRRB Publishing, ISBN-1549992406.

Stakem, Patrick H. *Rocketplanes to Space*, 2017, PRRB Publishing, ISBN-1549992589.

Stakem, Patrick H. *Crewed Space Stations,* 2017, PRRB Publishing, ISBN-1549992228.

Stakem, Patrick H. *Enviro-bots for STEM: Using Robotics in the pre-K to 12 Curricula, A Resource Guide for Educators,* 2017, PRRB Publishing, ISBN-1549656619.

Stakem, Patrick H. *STEM-Sat, Using Cubesats in the pre-K to 12 Curricula, A Resource Guide for Educators*, 2017, ISBN-1549656376.

Stakem, Patrick H. *Lunar Orbital Platform-Gateway*, 2018, PRRB Publishing, ISBN-1980498628.

Stakem, Patrick H. *Embedded GPU's*, 2018, PRRB Publishing, ISBN- 1980476497.

Stakem, Patrick H. *Mobile Cloud Robotics*, 2018, PRRB Publishing, ISBN- 1980488088.

Stakem, Patrick H. *Extreme Environment Embedded Systems,* 2017, PRRB Publishing, ISBN-1520215967.

Stakem, Patrick H. *What's the Worst, Volume-2*, 2018, ISBN-1981005579.

Stakem, Patrick H., *Spaceports*, 2018, ISBN-1981022287.

Stakem, Patrick H., *Space Launch Vehicles*, 2018, ISBN-1983071773.

Stakem, Patrick H. *Mars*, 2018, ISBN-1983116902.

Stakem, Patrick H. *X-86, 40<sup>th</sup> Anniversary ed*, 2018, ISBN-1983189405.

Stakem, Patrick H. *Lunar Orbital Platform-Gateway*, 2018, PRRB Publishing, ISBN-1980498628.

Stakem, Patrick H. *Space Weather*, 2018, ISBN-1723904023.

Stakem, Patrick H. *STEM-Engineering Process*, 2017, ISBN-1983196517.

Stakem, Patrick H. *Space Telescopes,* 2018, PRRB Publishing, ISBN-1728728568.

Stakem, Patrick H. *Exoplanets*, 2018, PRRB Publishing, ISBN-9781731385055.

Stakem, Patrick H. *Planetary Defense*, 2018, PRRB Publishing, ISBN-9781731001207.

Patrick H. Stakem *Exploration of the Asteroid Belt*, 2018, PRRB Publishing, ISBN-1731049846.

Patrick H. Stakem *Terraforming*, 2018, PRRB Publishing, ISBN-1790308100.

Patrick H. Stakem, *Martian Railroad,* 2019, PRRB Publishing, ISBN-1794488243.

Patrick H. Stakem, *Exoplanets,* 2019, PRRB Publishing, ISBN-1731385056.

Patrick H. Stakem, *Exploiting the Moon,* 2019, PRRB Publishing, ISBN-1091057850.

Patrick H. Stakem, *RISC-V, an Open Source Solution for Space Flight Computers,* 2019, PRRB Publishing, ISBN-1796434388.

Patrick H. Stakem, *Arm in Space*, 2019, PRRB Publishing, ISBN-9781099789137.

Patrick H. Stakem, *Extraterrestrial Life*, 2019, PRRB Publishing, ISBN-978-1072072188.

Patrick H. Stakem, *Space Command*, 2019, PRRB Publishing, ISBN-978-1693005398.

CubeRovers, A Synergy of Technologys, 2020, PRRB Publishing, ISBN-979-8651773138.

Robotic Exploration of the Icy moons of the Gas Giants. 2020, PRRB Publishing, ISBN- 979-8621431006

Hacking Cubesats, 2020, PRRB Publishing, ISBN-979-8623458964.

History & Future of Cubesats, PRRB Publishing, ISBN-979-8649179386.

Hacking Cubesats, Cybersecurity in Space, 2020, PRRB Publishing, ISBN-979-8623458964.

Powerships, Powerbarges, Floating Wind Farms: electricity when and where you need it, 2021, PRRB Publishing, ISBN-979-8716199477.

Hospital Ships, Trains, and Aircraft, 2020, PRRB Publishing, ISBN-979-8642944349.

<u>2020/2021 Releases</u>

*CubeRovers, a Synergy of Technologys*, 2020, ISBN-979-8651773138

*Exploration of Lunar & Martian Lava Tubes by Cube-X*, ISBN-979-8621435325.

*Robotic Exploration of the Icy moons of the Gas Giants*, ISBN- 979-8621431006.

*History & Future of Cubesats*, ISBN-978-1986536356.

*Robotic Exploration of the Icy Moons of the Ice Giants, by Swarms of Cubesats,* ISBN-979-8621431006.

*Swarm Robotics,* ISBN-979-8534505948.

*Introduction to Electric Power Systems*, ISBN-979-8519208727.

*Centros de Control: Operaciones en Satélites del Estándar CubeSat* (Spanish Edition), 2021, ISBN-979-8510113068.

*Exploration of Venus*, 2022, ISBN-979-8484416110.

Patrick H. Stakem, *The Search for Extraterrestial Life,* 2019, PRRB Publishing, ISBN-1072072181.

*The Artemis Missions, Return to the Moon, and on to Mars,* 2021, ISBN-979-8490532361.

*James Webb Space Telescope. A New Era in Astronomy*, 2021, ISBN-979-8773857969.

www.ingramcontent.com/pod-product-compliance
Lightning Source LLC
Chambersburg PA
CBHW020929180526
45163CB00007B/2950